OXFORD

in association with

science
museum

flight

OXFORD
UNIVERSITY PRESS

Oxford New York

Auckland Bangkok Buenos Aires Cape Town Chennai
Dar es Salaam Delhi Hong Kong Istanbul Karachi Kolkata
Kuala Lumpur Madrid Melbourne Mexico City Mumbai Nairobi
São Paulo Shanghai Taipei Tokyo Toronto

Published by Oxford University Press, Inc.
198 Madison Avenue, New York, NY 10016
www.oup.com

Oxford is a registered trademark of Oxford University Press

Text © Philip Wilkinson 2003

Database right Oxford University Press (maker)

First published in 2003

Library of Congress Cataloging-in-Publication Data is available.

ISBN 0–19–521996–1

1 3 5 7 9 10 8 6 4 2

Printed in Italy

Acknowledgments

The publishers would like to thank:
For the Science Museum: Peter Davison

All photos reproduced in kind permission of the Science and Society Picture Library
with the exception of the following:
Airborne (AUS); p19bl
Breitling SA © Cameron Balloons; p9bl (Jean-Francois Luy)
Digital Vision; p14bl, 16-17c, 17r, 20-21
NASA; p7br

OXFORD
in association with

science
museum

flight

Contents

Age of the birdmen

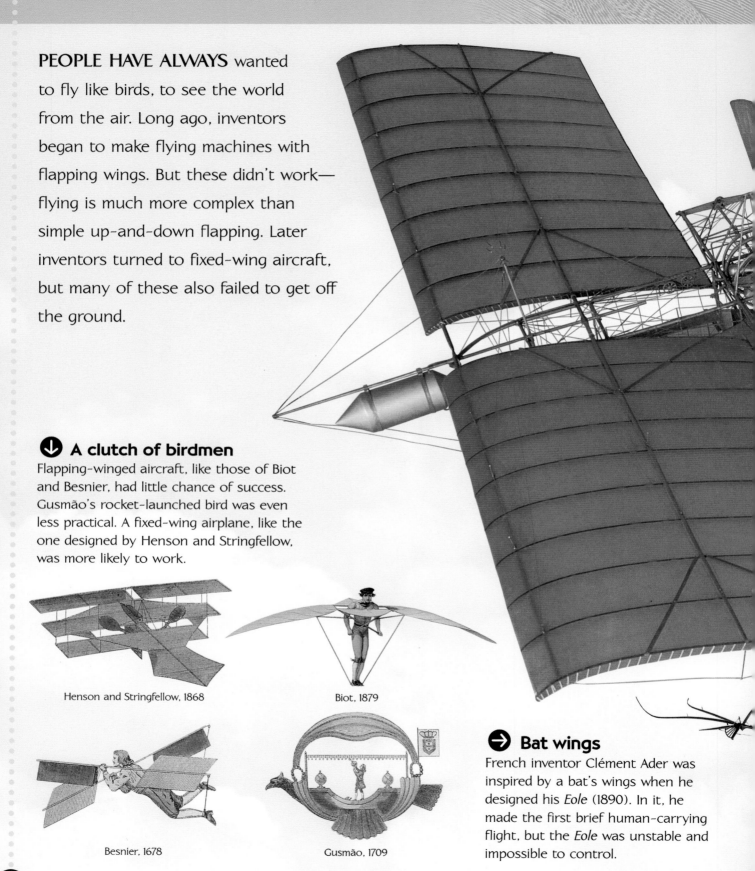

PEOPLE HAVE ALWAYS wanted to fly like birds, to see the world from the air. Long ago, inventors began to make flying machines with flapping wings. But these didn't work—flying is much more complex than simple up-and-down flapping. Later inventors turned to fixed-wing aircraft, but many of these also failed to get off the ground.

⬇ A clutch of birdmen
Flapping-winged aircraft, like those of Biot and Besnier, had little chance of success. Gusmão's rocket-launched bird was even less practical. A fixed-wing airplane, like the one designed by Henson and Stringfellow, was more likely to work.

Henson and Stringfellow, 1868

Biot, 1879

Besnier, 1678

Gusmão, 1709

➡ Bat wings
French inventor Clément Ader was inspired by a bat's wings when he designed his *Eole* (1890). In it, he made the first brief human-carrying flight, but the *Eole* was unstable and impossible to control.

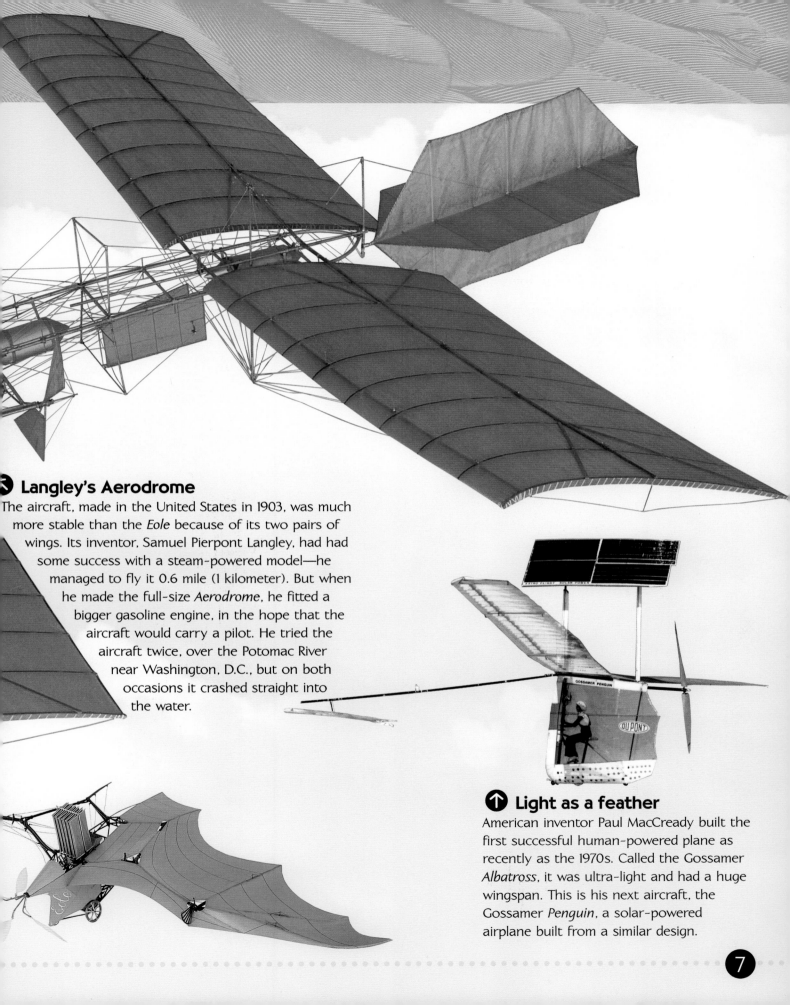

Langley's Aerodrome

The aircraft, made in the United States in 1903, was much more stable than the *Eole* because of its two pairs of wings. Its inventor, Samuel Pierpont Langley, had had some success with a steam-powered model—he managed to fly it 0.6 mile (1 kilometer). But when he made the full-size *Aerodrome*, he fitted a bigger gasoline engine, in the hope that the aircraft would carry a pilot. He tried the aircraft twice, over the Potomac River near Washington, D.C., but on both occasions it crashed straight into the water.

Light as a feather

American inventor Paul MacCready built the first successful human-powered plane as recently as the 1970s. Called the Gossamer *Albatross*, it was ultra-light and had a huge wingspan. This is his next aircraft, the Gossamer *Penguin*, a solar-powered airplane built from a similar design.

Floating through the air

BALLOONS ARE SO LIGHT that they simply float up into the air. To make a balloon light enough, you need to fill it with a gas that is less dense than the air around it. The earliest balloons, which were the first successful flying machines, were filled with hot air. Many modern balloons still use hot air, supplied by a gas burner, but other gases, such as hydrogen or helium, can also be used.

➡ First up

In 1783, the French Montgolfier brothers filled an enormous, elaborately decorated balloon with hot air and launched it over Paris with two of their friends, François Pilâtre de Rozier and the Marquis d'Arlandes, aboard. The flight was the first to carry people and was a huge success. It lasted about 25 minutes and covered roughly 6.2 miles (10 kilometers), at a relatively low altitude. When they landed, the two men were mobbed by an excited crowd who tore Pilâtre's jacket to pieces for souvenirs.

⬅ Flying in a basket

Shortly after the Montgolfiers, two other French pioneers, the Robert brothers, took off on the first flight in a hydrogen balloon. Their balloon was designed by French scientist Jacques Charles. It had many features used in later balloons, such as a clever valve to let out gas, so that the balloon could descend.

⬆ Ship in the air

Airships, like this British military one from the 1920s, are rather like large, cigar-shaped balloons. But, unlike a balloon, an airship has an engine and can be steered.

⬊ World voyager

In March 1999 Bertrand Piccard and Brian Jones were the first to fly a balloon nonstop around the world. Climbing to a staggering 38,566 feet (11,755 meters) above the Earth's surface, they traveled in a sealed capsule beneath their huge balloon, Breitling Orbiter 3, which was filled with helium and hot air.

Airplane pioneers

IN THE EARLY YEARS of the 20th century every airplane was experimental, and every flight was a dangerous adventure. The aircraft were light, fragile, and difficult to fly. But, as pilots and designers gained experience, more and more airplanes got off the ground.

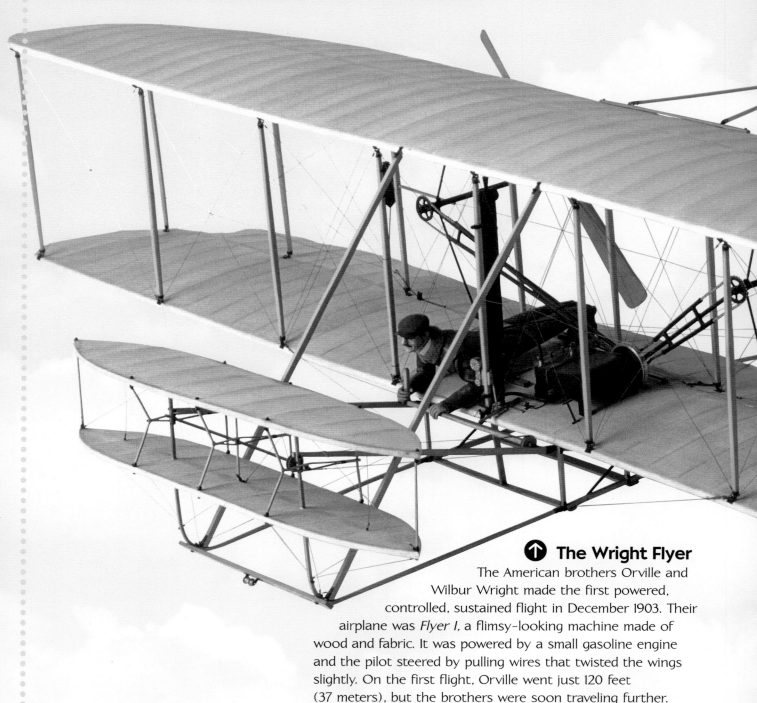

⬆ The Wright Flyer
The American brothers Orville and Wilbur Wright made the first powered, controlled, sustained flight in December 1903. Their airplane was *Flyer I*, a flimsy-looking machine made of wood and fabric. It was powered by a small gasoline engine and the pilot steered by pulling wires that twisted the wings slightly. On the first flight, Orville went just 120 feet (37 meters), but the brothers were soon traveling further.

↑ Channel crossing

The first aviator to cross the English Channel was Frenchman Louis Blériot in 1909. He became instantly famous, and soon many people were buying airplanes of his design.

↗ Come to the show

Everyone was interested in airplanes, and soon pilots were thrilling the crowds at air shows. For brave spectators, there might even be the chance of a "spin" as a passenger.

↗ How many wings?

Early single-winged craft tended to be rather weak and prone to accidents, so many pilots favored two or three sets of wings. Triplanes, like this German Fokker from World War I, were strong and their triple wings gave good lift, but wind resistance slowed them down.

Streamlined speed

IN THE 1930s, airplanes were transformed. Smooth streamlined metal bodies and more powerful engines helped them go much faster than before. These aircraft also had closed cabins that could carry more people in comfort. Suddenly, everyone who could afford it wanted to travel by air.

⬇ Piggyback plane

Take-off could be a problem for a heavily loaded aircraft, because it needed more power and therefore heavier fuel. British designer Robert Mayo came up with one answer—the composite, or "piggyback," airplane. This was actually two planes in one, a large flying boat and a smaller seaplane. The two aircraft took off together, using the power of the flying boat's big engines. Then the seaplane was disconnected, and could fly for around 6,000 miles on its small engines.

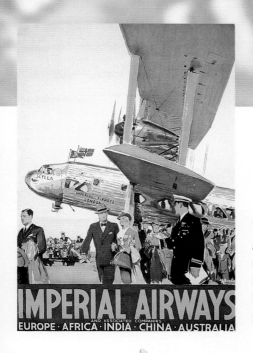

← All aboard

Many airports were built as major airlines opened for business in the 1930s. Britain's Imperial Airways flew all over the world.

↓ Fast flyer

The Boeing 247D could carry passengers in reasonable comfort at up to about 186 mph (300 km/h). But it could only hold 10 people, so was not very popular with the airlines.

↓ Worldwide success

The Douglas DC3 of 1936 was the most successful airliner of all time. It could carry up to 21 people quickly and cheaply, and was bought in large numbers by airlines all over the world. These reliable aircraft are still used in many places today.

Shrinking the world

JET AIRCRAFT were very different from earlier planes. Jets are amazingly fast, but they also fly well above the weather, making the journey smooth. They can also be large, carrying lots of passengers. By the late 1950s jet-engined airliners were speeding people all over the world. For many travelers, the world really did seem a smaller place.

◣ Pioneer jet

Introduced in 1952, the De Havilland Comet was the world's first jet airliner. Passengers liked this sleek, fast plane, but a structural weakness led to several accidents, so it had to be redesigned.

↘ What's inside?

A cutaway model of a Vickers Vanguard airliner shows how the baggage is stowed neatly in the hold, below the passengers' feet. The wings contain the aircraft's fuel tanks.

↓ Record-breaker

Concorde is by far the world's fastest airliner: it can travel from London to New York—about 3,500 miles (5,600 kilometers)—in 3 hours 30 minutes, at over twice the speed of sound. Its sleek shape, with narrow fuselage and streamlined nose, helps it to achieve this supersonic success. During landing the nose hinges downward, so the pilot can see the ground more clearly. When first introduced in 1976, Concorde was criticized for being too noisy. In spite of this, and a fatal accident in 2000 that led to some design changes, Concorde was a highly successful airliner.

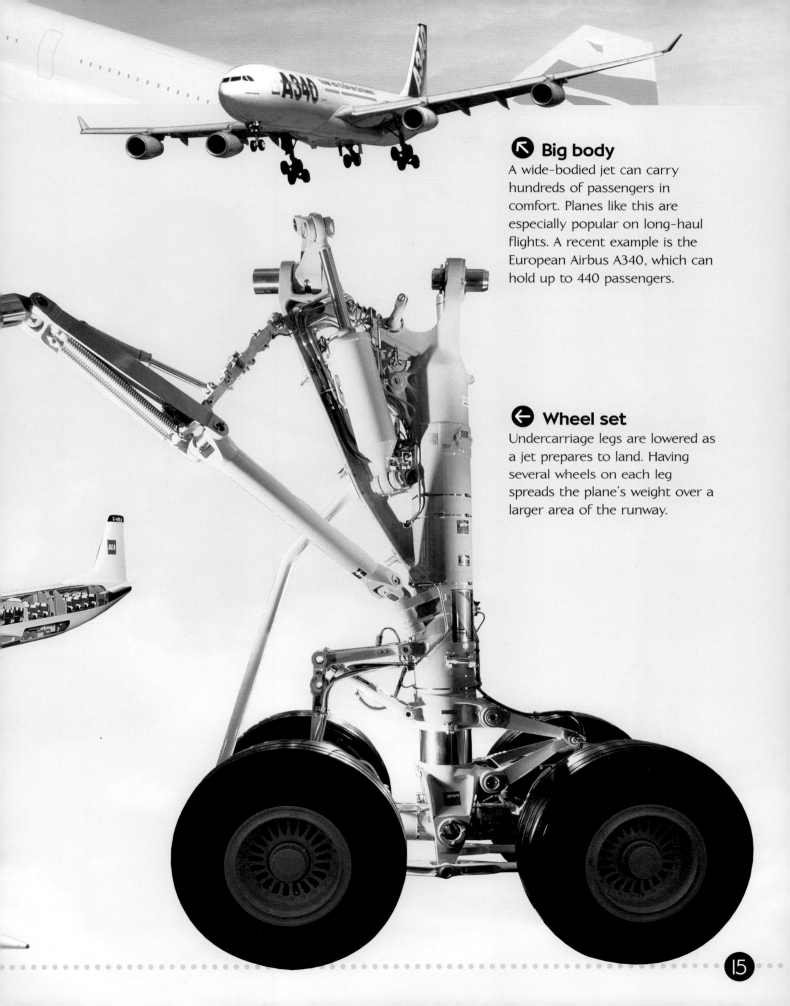

Big body

A wide-bodied jet can carry hundreds of passengers in comfort. Planes like this are especially popular on long-haul flights. A recent example is the European Airbus A340, which can hold up to 440 passengers.

Wheel set

Undercarriage legs are lowered as a jet prepares to land. Having several wheels on each leg spreads the plane's weight over a larger area of the runway.

Whirling blades

WHIRLING AROUND at terrific speed, a helicopter's rotor blades act like a wing and propeller rolled into one. The blades pull the helicopter up into the air, so that it can take off vertically. Rotor blades also make helicopters very maneuverable—they can fly very slowly, hover, and land in a confined space. Helicopters are used for all sorts of jobs, from filming to police work, where these qualities are needed.

⊽ Artist's vision
Italian artist and inventor Leonardo da Vinci drew this wood-and-canvas rotorcraft in the 15th century. This artist's dream never flew.

⊾ Spin and lift
Designed in the 1930s by Spanish engineer Juan de la Cierva, the autogiro is a cross between an airplane and a helicopter. The power comes from the propeller, while the free-spinning rotor provides lift.

↙ Work horse

Nicknamed the "Sea Stallion," Sikorsky's CH-53E is a big three-engined helicopter. It is used in amphibious warfare, for carrying troops and heavy equipment. It often operates from aircraft carriers and assault ships. It can be fitted with night-vision equipment, so that it can carry out low-level work in the dark.

↑ To the rescue

Helicopters such as this Sea King are the perfect air-sea rescue aircraft, because they can hover while an injured or stranded person is winched on board.

← All-round vision

Many helicopters, like this Sikorsky, have windows all around the cockpit. This makes them perfect for all sorts of jobs, from military observation to checking up on road traffic, where it is vital to be able to see well.

Aircraft for all

MANY LIGHT AIRCRAFT—especially the tiny, cloth-winged microlights—look so small and delicate that it seems amazing they can fly at all. Yet there are thousands of small aircraft in the skies, flown for work, for fun, for transport, or to provide breathtaking aerobatic displays.

➡ Going for a spin

Spinning through the sky, as comfortable flying upside down as the right way up, the Pitts Special is the ultimate aerobatic plane. The first Specials were designed by American enthusiast Curtis H. Pitts in the 1940s. Their biplane structure makes them both light and strong, and they are also very agile—a Pitts Special can do maneuvers that pilots of other aircraft can only dream of. As a result, the Special has won many aerobatics competitions, and many are still being flown today.

⬇ In control

The delicate-looking Pathfinder microlight has many of the features of a larger plane. Its wings and tail have control surfaces (movable flaps) that enable the pilot to steer.

⬆ Bumble Bee

The Z37 Cmelak "Bumble Bee" is a strong, lightweight, closed-cockpit airplane used for crop-spraying and other agricultural work. It was built in 1967 in Czechoslovakia (now the Czech Republic).

← Flexible wing

This twin-seater microlight looks rather like a kite. The pilot steers by moving a metal bar linked to the fabric wing, which bends slightly to turn the plane. Hold on tight!

War in the air

AIRPLANES have been used by the military since World War I. They play a key role in modern warfare, especially in reconnaissance (spying on the enemy) and in bombing. Designers use the most up-to-date technology to aim bombs with pinpoint accuracy, and to protect pilots and crew from enemy fire.

➔ Big bird

The Blackbird was developed as an ultra-fast spy plane. It can travel at three times the speed of sound and flies very high, at over 80,000 feet (24,384 meters). At such speeds and heights, it has to withstand much more stress and strain, and far higher temperatures, than an ordinary aircraft. For this reason its designers used special, high-strength materials. They also had to design most of the plane's equipment and fittings from scratch— the Blackbird has its own special tires, fuel, and even paint.

⬅ Swarm of Hornets

The F/A-18 Hornet is used in several roles, from firing missiles to dropping bombs. It is designed for use on aircraft carriers, but is also very fast, up to 1.8 times the speed of sound.

➡ Super stealth

The unusual shape and special black coating of the American B2 Spirit stealth bomber make it virtually invisible to radar, as it darts through the air at up to 645 mph (1,038 km/h).

Glossary

aerobatics Art of performing daring maneuvers in an aircraft, often for entertainment.

aircraft carrier Ship from which aircraft can take off and on which they can land.

airline A company that carries passengers or goods by air to specified destinations.

airliner Large passenger-carrying aircraft.

aviator Pilot of an aircraft; this term is used especially to refer to the pioneer pilots in the early years of flight.

cabin Section of an aircraft used to carry the passengers and their hand luggage.

fixed-wing aircraft Airplane with rigid wings that cannot move or flap (as opposed to rotary-winged aircraft such as helicopters).

flying boat Large seaplane with a boat-like body.

helium Gas sometimes used in balloons and airships; it is very light and also inert, which makes it much safer than hydrogen.

hover To remain in the air without moving forward or backward, up or down.

hydrogen Gas sometimes used in balloons and airships; it is the lightest of all gases, but also catches fire very easily.

light aircraft Small airplane, used for carrying people over short distances, for pleasure flights, pilot training, or aerobatics.

rotor Set of spinning blades fitted to a helicopter or autogiro; on a helicopter, the rotor provides both lift and power, on an autogiro the unpowered rotor provides only lift.

seaplane Ordinary airplane fitted with floats instead of wheels, so that it can land on water.

streamlined Term used to describe an aircraft, boat, or vehicle designed with a smooth shape so that it moves more freely through air or water.

triplane Airplane with three sets of wings.

undercarriage The landing gear of an aircraft, used during take-off and landing and consisting of a set of wheels on a supporting framework; on many airplanes, the undercarriage can be raised or lowered during flight.

valve Device that can be used to control the flow of a liquid or gas, for example, for letting the gas out of a balloon.

wingspan The distance from one tip of an aircraft's wing to the other wing-tip.

Index